MAR -- 2021

# LEGO NINJAGO

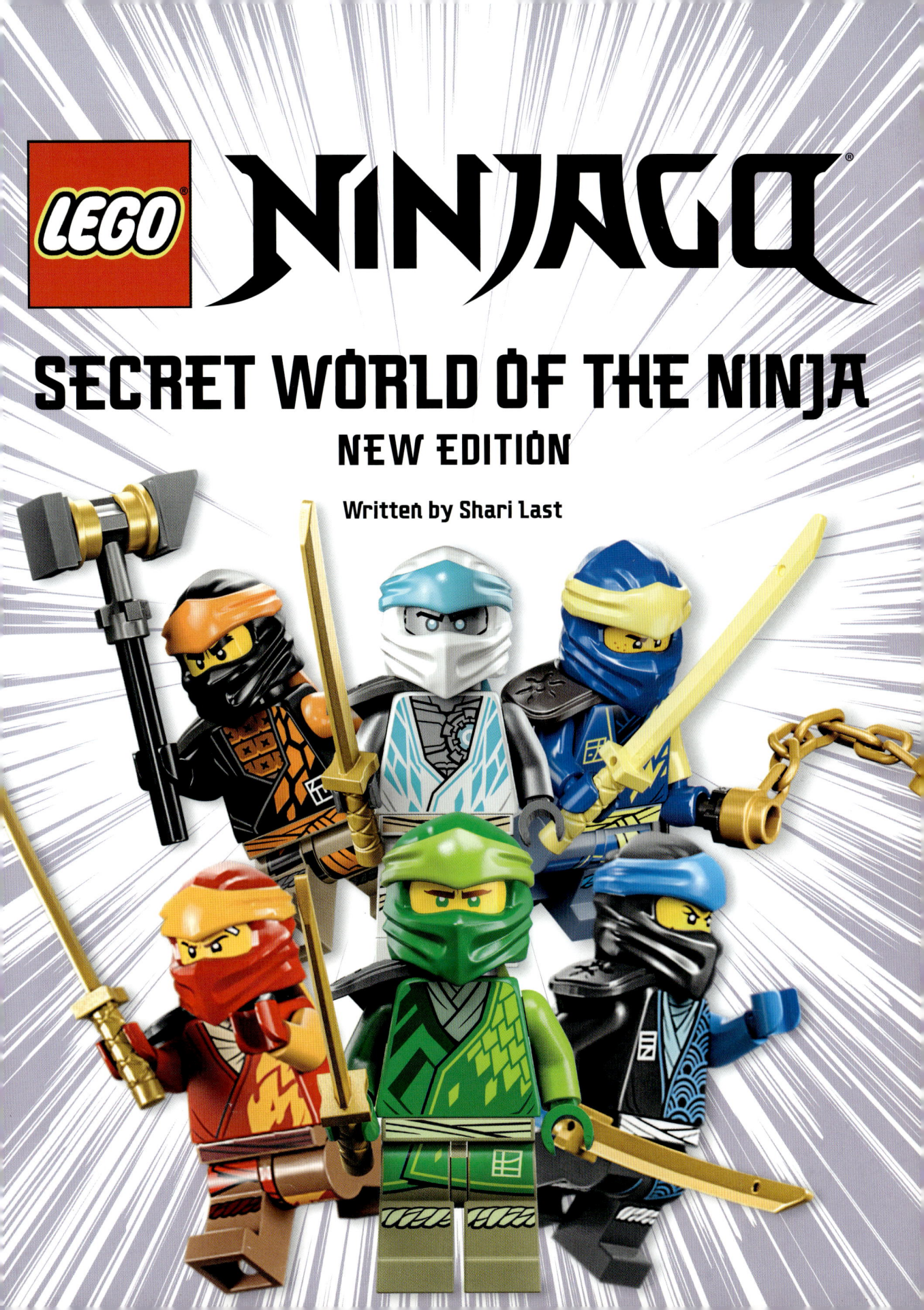

## SECRET WORLD OF THE NINJA
### NEW EDITION

Written by Shari Last

# Contents

| | |
|---|---|
| Introduction | 4 |
| Wu | 6 |
| A Ninja Is … | 8 |
| Kai | 10 |
| So You Want to Be a Ninja? | 12 |
| Garmadon | 14 |
| Ultra Dragon Attack! | 16 |
| What Do Ninja Wear? | 18 |
| Jay | 20 |
| Where Do Ninja Live? | 22 |
| Vile Villains | 24 |
| Lloyd | 26 |
| Dragons | 28 |
| How Do You Find Your Dragon Form? | 30 |
| Myths and Legends | 32 |
| Nya | 34 |
| Battle in the Deep | 36 |
| Overlord | 38 |
| Enemies of Ninjago | 40 |
| Zane | 42 |
| Fire Stone Mech Escape | 44 |
| Weapons | 46 |
| Cole | 48 |

| | |
|---|---|
| Ultra Sonic Raider | 50 |
| Mastering the Elements | 52 |
| Who Will Win the Tournament of Elements? | 54 |
| Good and Evil | 56 |
| Ice Dragon Attack | 58 |
| Skylor | 60 |
| Ninja Gamers | 62 |
| P.I.X.A.L. | 64 |
| How Do Ninja Train? | 66 |
| Princess Harumi | 68 |
| Can the Ninja Stop the Sons of Garmadon? | 70 |
| Teamwork | 72 |
| Battle at Sea! | 74 |
| Armies Attack! | 76 |
| Escape From the Dungeons of Shintaro | 78 |
| What Are the Fundamental Ninja Virtues? | 80 |
| Monastery of Spinjitzu | 82 |
| Crystal King | 84 |
| The Final Battle | 86 |
| Attack on the Crystal Palace | 88 |
| A New Reality | 90 |
| Can New Heroes Join Team Ninja? | 92 |
| Glossary | 94 |
| Index | 95 |
| Acknowledgments | 96 |

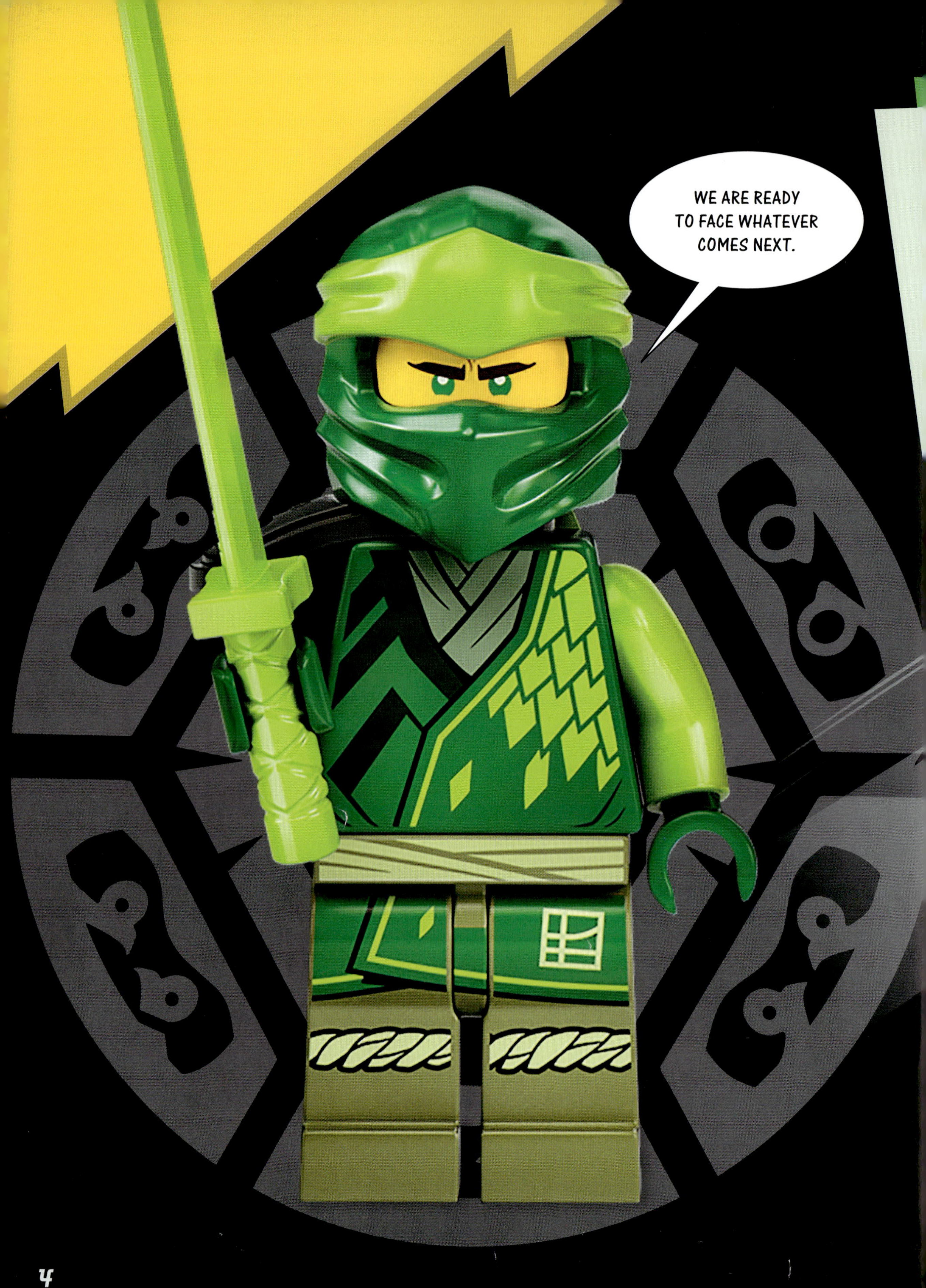
WE ARE READY TO FACE WHATEVER COMES NEXT.

# Introduction

Welcome to Ninjago, a place of wonder.

A team of Ninja warriors defends its realm against all forms of evil. And, as the Ninja quickly learn, evil has many, many forms. The Ninja harness the power of the elements. They master ancient weapons, unearth legendary artifacts, and unleash mighty dragons—and there are always more secrets to be discovered.

Fighting evil used to be a lot simpler. Now, strange new stories are being whispered across Ninjago. Tales of dragons, forgotten legends, and mysterious new threats. But Team Ninja is ready. The Ninja are trained to adapt. They will never stop learning. They will never give up!

## Baby Wu

It's been a long time since Wu was a baby, but after an encounter with the Reversal Time Blade, the Ninja get to meet Baby Wu. He enjoys silly jokes, being cuddled by Cole, and drinking tea, of course.

*I'M WISE BUT COOL!*

Symbol of the Firstbourne

Dragonbone Blade

## Dragon Master

Wu understands that power comes from within. He wears the ancient Dragon Armor—but it is his purity of heart that makes him the true Dragon Master.

### DATA FILE

- **Also known as:** Lloyd's uncle
- **Known for:** Unusual sense of humor
- **Likes:** Dancing and tea
- **Dislikes:** Students ignoring his advice

# Wu

**SON OF THE** First Spinjitzu Master, wise Wu follows in his father's footsteps and does all he can to protect Ninjago. He can harness the mighty power of Creation, but prefers to put all his energies into training Team Ninja. In times of trouble, Wu can be counted on for help, good advice, and bad jokes.

Trademark white beard

### DID YOU KNOW?
Wu has been turned into an evil cyborg, trapped in the Sword of Souls, and stranded in another realm.

## Top Teacher
Wu is a skilled Ninja who knows many of Ninjago's deepest secrets. But his greatest strength lies in teaching. He loves inspiring and motivating his students.

Wooden shoulder pad

Ivory Blade of Deliverance

## Warrior Wu
When the Ninja need him, Wu is always battle ready. He puts the combat skills he learned from his father to good use, defeating villains and protecting his friends.

# A Ninja Is ...

**EVERY NINJA IS** different, but they have certain qualities in common. Ninja are brave. Ninja are skilled. Ninja are disciplined. Ninja are good at making tea for Master Wu. But what else makes a Ninja? Let's hear from some of the Ninja's best friends and worst enemies.

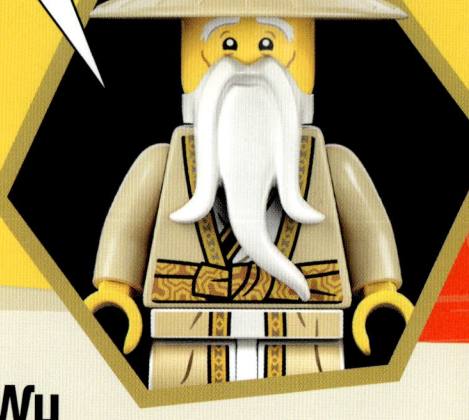

> A NINJA IS A MASTER OF HIS OR HER OWN TALENTS.

## Wu
One of the Ninja's biggest supporters, Master Wu taught them everything he knows. He is proud of his former students and can't wait to see the great things they will achieve in the future.

## Vangelis
Vangelis was thwarted twice by the Ninja, once as the terrifying Skull Sorcerer and again as a member of the Crystal Council. It's safe to say he isn't the biggest Ninja fan.

> A NINJA SHOULD BE STEALTHY— BUT QUICK.

> NINJA CAN'T KEEP THEIR NOSES OUT OF OTHER PEOPLE'S BUSINESS!

## Griffin Turner
Master of Speed and an old ally of the Ninja, Griffin even became a Ninja for a little while. He helped save Ninjago from the stone giant, Colossus.

# Mechanic

The Mechanic isn't the worst villain in Ninjago, but he isn't the nicest guy either. He was not impressed when the Ninja captured him and stole his identity to help save Ninjago City.

*THOSE NINJA ARE NOT TO BE TRUSTED.*

*NINJA WILL RUN INTO DANGER TO PROTECT THEIR FRIENDS.*

# Skylor

Skylor can copy the elemental powers of others, so she knows better than most what it takes to be a Ninja. She has teamed up with the Ninja on several dangerous occasions.

# Pythor

This sneaky snake has never met a Ninja he liked. For some reason, they always try to stop him from carrying out his evil plans.

*ONE DAY I'LL SSSSTOP THOSE NINJAS!*

# Jet Jack

Now an ally of the Ninja, Jet Jack didn't always feel this way. She once led the Dragon Hunters to capture the Ninja—but it didn't take long for her to figure out who the good ones really were.

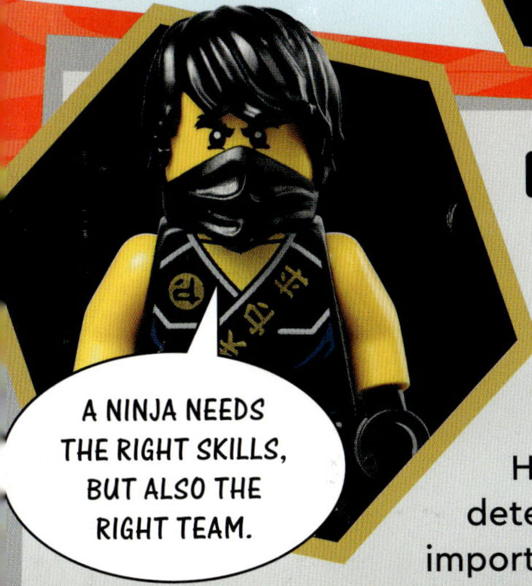

*A NINJA NEEDS THE RIGHT SKILLS, BUT ALSO THE RIGHT TEAM.*

# Cole

One of the original members of Team Ninja, Cole knows exactly what it takes to be one. Hard work and determination are important, but so is the ability to work as a team and put up with endless Ninja pranks.

*A NINJA ALWAYS FIGHTS FOR WHAT'S RIGHT.*

## Old Skills

Before Kai's Ninja days he was a blacksmith, forging tools in the heat of a raging fire. When the Oni invade Ninjago, Kai calls on his old skills to reforge the Golden Weapons.

BRING ON THE HEAT!

## Protective Powers

When the Crystal King and his army threaten Ninjago, Kai and his fellow Ninja sacrifice their own skills to save others. This selfless act unlocks their Dragon Form and gives them golden, glowing wings.

### DATA FILE

- **Also known as:** Master of Fire
- **Known for:** Loyalty to his friends
- **Likes:** Skylor
- **Dislikes:** Gingerbread men

# Kai

**KAI WAS THE LAST** to join the original Team Ninja, but he is always the first to leap into battle. He is fiery and stubborn by nature, which often gets him into trouble. But Kai's stubbornness is also his greatest asset because it helps him get the job done no matter what.

## Family Matters
Kai and his sister, Nya, reunite with their parents, who disappeared years ago. Ray and Maya are the former Elemental Masters of Fire and Water.

Golden Dragon blades

Kai and Skylor inside cockpit

Armored sides

### DID YOU KNOW?
Kai once opened his own Ninja training dojo, where he taught Ninja skills to young kids.

Wheels fold back for walker mode

## Red Raider
Kai's Golden Dragon Raider was created by P.I.X.A.L. and specially engineered to withstand the power-draining abilities of Vengestone. At night, its red coloring changes to black for better camouflage.

# So You Want to Be a Ninja?

> LET'S START WITH THE NINJA ESSENTIALS. POWERS AND DRAGONS CAN WAIT!

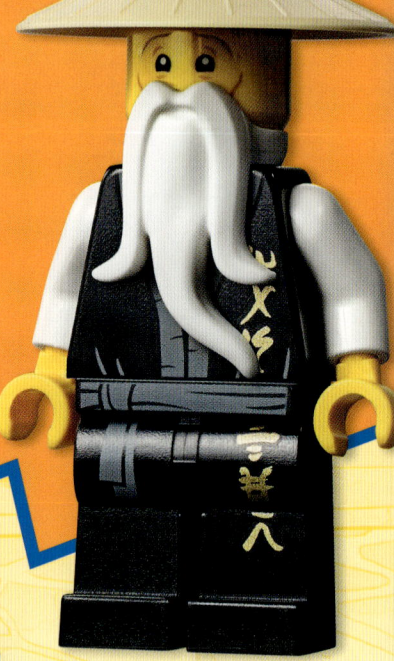

**BECOMING A NINJA** might be the coolest thing ever, but it doesn't happen overnight. Master Wu is here to teach the Ninja all sorts of awesome skills, including how to harness their elemental powers. But the first thing he teaches them is that some things are even more important than powers.

## Patience

Most of the time, the Ninja want to jump right into their training, right onto their dragons, or right up into a burst of Airjitzu. Master Wu teaches them to slow down, preferably with a cup of tea. Patience is what will help the Ninja train well, stay calm in battle, and make smart decisions.

# Mastery

The Ninja are excited about becoming Elemental Masters, but quickly realize it will take a lot of practice and training. Lloyd struggles to master the power of Energy at first, but he realizes that true mastery is worth the wait.

# Teamwork

Training, working, and living together means there are moments when the Ninja don't see eye to eye. But when it comes to taking on their enemies, all quibbles are forgotten. The Ninja are far more powerful as a team than they are on their own.

# Focus

A Ninja must be able to focus on the task at hand—and not get caught up in other pursuits along the way. Master Wu teaches the Ninja this valuable lesson by setting them an obstacle course full of distractions.

"I'M NOT ALL BAD—PROMISE!"

## Brothers at War

Garmadon and his brother, Wu, are total opposites. They have fought against each other many times, though sometimes they find themselves fighting for the same side.

Spinjitzu harnesses the power of Destruction

## Father and Son

Garmadon cares for his son, Lloyd, more than anything. While it's true that Garmadon has tried to fight Lloyd at times, the two are developing a strong bond as time goes on.

### DATA FILE

- **Also known as:** King of Shadows
- **Known for:** Having a bad side
- **Likes:** Gardening
- **Dislikes:** Losing a fight

# Garmadon

GARMADON IS NO ordinary villain. He was once an evil Lord and a cruel Emperor, but he's also a loving father, Ninja Master, and—at times—a brave hero! Garmadon tries to overcome his evil nature, but it's hard sometimes.

*Face looks like an Oni mask*

### DID YOU KNOW?
Garmadon wasn't always so complicated. As a child, he was bitten by the Great Devourer, which turned him evil.

## Oni Form
Garmadon is part Oni, a tribe of demons that spreads darkness and destruction. His Oni form has altered his appearance permanently, and he often calls on his Oni powers during battle.

*Shaped like the ancient Helmet of Shadows*

*Treads flatten everything in its path*

## The Garmatron
Garmadon's terrifying tank shoots Dark Matter, a mysterious substance that turns people evil. But Garmadon doesn't realize that upsetting the balance between light and dark will allow the Overlord to return to Ninjago. Oops!

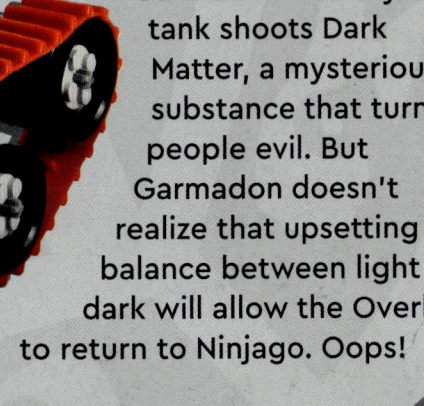

15

# Ultra Dragon Attack!

**WHEN THE GREAT DEVOURER** threatens to consume the whole of Ninjago, the Ninja merge their four Elemental Dragons to form the amazing Ultra Dragon. It soars through the skies toward the giant snake, while everyone hopes its four heads will be better than one!

Mouth channels the might of Earth

### Dinner Time!
The Great Devourer is a gigantic snake. Its poison bite can turn a person evil—which is exactly what happened to Lord Garmadon as a boy. If left to its own devices, this monster will consume all of creation!

Icy breath freezes whatever it touches

**BATTLED ALONGSIDE THE NINJA 4 TIMES**

### DID YOU KNOW?
As the Green Ninja, Lloyd is the only one who can ride the Ultra Dragon. He gives his new pet a very original nickname: Ultra.

Wings large enough to support four-headed body

Seat allows for a more comfortable ride

Deadly whipping tail

Mouth spits fireballs

## Power Up
The Ultra Dragon combines the four Elements of Creation: Fire, Earth, Ice, and Lightning. Each of its four heads harnesses an elemental power—which makes this dragon very powerful indeed.

# What Do Ninja Wear?

**NINJA GEAR IS CONSTANTLY** adapted or upgraded for various battles in different environments. The Ninja rely on their energetic Spinjitzu moves, so they have three requirements when it comes to their robes—they should be comfortable and flexible, and they need to look cool!

### Prison Grays
Spending time in Kryptarium Prison is not fun—and the itchy uniform makes it even worse. Lloyd can't believe how uncool the striped shirts are!

### Casual Clothes
A Ninja needs to relax sometimes, too. When Nya isn't performing her Ninja duties or battling as Samurai X, she dresses like any other Ninjago citizen.

### Golden Ninja
When Lloyd uses his Golden Power, his Ninja suit turns to gold. It keeps some green details, though. He *is* the Green Ninja after all!

### Airjitzu Robes
The Ninja earn these robes after unlocking this new skill. Each robe displays the Airjitzu symbol, as well as an animal unique to each Ninja. These suits are also known as Destiny Robes.

## Dojo Suit
These simple suits are perfect for grueling dojo training. A bold pattern of the Ninja's Elemental Power runs across the top, and the Ninja's initial is marked on the bottom corner.

## Island Gear
Equipped for exploring unknown locations, these suits have utility belts, straps, and pockets to hold gadgets, rations, and anything else that might come in handy!

## Hero Armor
The Ninja receive their gold or silver Hero armor in Shintaro. It is engraved with their Elemental symbols and fits over their robes. There is a shoulder piece on one side.

## Pink Edition
Ninja should remain focused at all times—even when doing the laundry. Kai's red robes once found their way into the whites, leaving Zane's training robes a lovely pink.

## Spinjitzu Burst
When a Ninja learns how to perform the Spinjitzu Burst, their elemental energy flows through their body and even blazes out of their eyes! Cole is the first Ninja to learn this move.

## Tea Shop T-Shirts
The Ninja weren't too impressed when Wu asked them to advertise his tea shop, Steep Wisdom. But out of respect for their Master, they wore the goofy uniforms over their armor!

## DID YOU KNOW?
The words or letters on the Ninja's robes are written in the Ninjago alphabet.

19

> MY LIGHTNING MOVES ARE NO JOKE!

## Lighthouse
When Jay thinks his girlfriend, Nya, is lost, his happy, jokey side disappears. He hides away in a remote lighthouse where he collects water and paints sad self-portraits.

Space helmet gives Jay a customized in-game look

Weapons bought with game credits

Energy monitor

## Gamer
No one loves video games more than Jay, so he can't believe his luck when he gets to enter one of his favorites, *Prime Empire*. Inside the game, Jay's avatar gets the nickname "Superstar Rockin' Jay." He can't pretend he doesn't love it!

### DATA FILE
**Also known as:** Master of Lightning

**Known for:** Electric sense of humor

**Likes:** Nya

**Dislikes:** Itchy fake beards

# Jay

**IT CAN BE HARD** to believe that this joker, gamer, and groovy dancer is also a fast-moving, quick-thinking Ninja. Jay is a core member of the team, coming up with ingenious plans, high-tech inventions, and much needed humor for when things get too dark.

Scuba helmet with rebreather

## Team Player
Jay makes a lot of silly jokes, but when it comes to protecting his friends, he is totally serious. After the Hydro Bounty ship breaks, Jay risks his life to power up the engine.

### DID YOU KNOW?
Jay and Nya have been through many dangerous adventures together. Recently, Jay asked Nya to be the Yang to his Yin.

## Lightning Skills
Jay's elemental skills include harnessing the power of lightning and powering up electronic devices. In his Dragon Form, Jay joins forces with the other Ninja to channel the power of Creation.

Dragon wings allow flight

Golden Dragon helmet

Golden Dragon motorcycle

21

# Where Do Ninja Live?

**MONASTERY HAS BEEN DESTROYED 2 TIMES**

**AFTER A HARD DAY'S** Ninja-ing, the team needs a home where the Ninja can train, sleep, and eat—and where Lloyd can play pranks on his friends! The Ninja have called many places home but, unfortunately, most of their bases keep getting demolished.

## Destiny's Bounty

No place has ever felt more like home than the Ninja's trusty ship. Not only does the Destiny's Bounty have a training room, weapons store, cozy kitchen, and TV room—it can also fly!

# Monastery

The original home of the Ninja, the Monastery of Spinjitzu, was a little on the small side. It has now been rebuilt with plenty of space for advanced Ninja training.

Master Wu loves taking a quiet moment in his tearoom at the Dojo Temple.

# Dojo Temple

When you're constantly defending your city, it pays to have a training arena, weapons room, and workshop close at hand. The secret Dojo Temple is equipped with everything a Ninja could ask for.

# Mobile Base

When the Ninja head to Master Chen's island, Nya brings along a very handy mobile base that she designed herself.

# Vile Villains

**ALL SORTS OF ENEMIES** threaten Ninjago—from ghosts and skeletons to criminals and sorcerers. Some villains have powers to rival the Ninja's, while others are simply greedy, cruel, or just plain mean! With new foes turning up all the time, the Ninja wonder who they will be up against next ...

### Samukai
King of the Underworld, Samukai leads the Skulkin in search of the four Golden Weapons. He thinks his four arms are perfect for holding all four weapons. But the Ninja won't make it easy!

### Master Chen
A noodle tycoon turned Anacondrai leader, Master Chen invited the Ninja to his island so he could steal their powers.

### Morro
The spirit of Wu's first Ninja student returns and possesses the body of Lloyd. Morro is searching for the powerful Realm Crystal to free his master.

### Nadakhan
Nadakhan is a mystical being known as a djinn. He's angry because he's spent the last 500 years trapped inside a teapot. Once free, he sets out to rebuild his realm, Djinjago.

### Time Twins
Twins Acronix and Krux are the Elemental Masters of Time. Or they were. Or they have been! They travel through time, intent on ruling Ninjago.

## Ultra Violet
Unpredictable and cruel, Ultra Violet is a general of the Sons of Garmadon. She chases after the Ninja and delights when things don't go their way.

## Iron Baron
The leader of the Dragon Hunters is on a dragon hunt! He seeks the Firstbourne, an ancient dragon that ate his arm and leg!

## Aspheera
An ancient Hypnobrai sorceress, Aspheera is fueled by the thought of revenge against Wu and his Ninja. Her magical staff is the source of her powers.

## Unagami
The artificial intelligence system behind the video game *Prime Empire*, Unagami is not satisfied with living inside a game. He decides it's time to break out into the real world of Ninjago.

## Mr. E
A Nindroid general with the Sons of Garmadon, Mr. E develops a fierce rivalry with Zane. He calls on the powers of an Oni mask to fight against the Ninja.

## Skull Sorcerer
This mysterious villain uses the powers of an ancient skull to create an army of resurrected warriors. When the Ninja try to stop him, they discover he is actually the King of Shintaro!

## DID YOU KNOW?
Many villains end up in Kryptarium prison, where they spend a lot of time talking about how much they dislike the Ninja!

"DIDN'T YOU KNOW? A NINJA NEVER QUITS!"

## Troubled Youth
Looking back now, it's hard to believe all the things Lloyd got up to when he was younger! He once tried to follow in his father's evil footsteps by trying to steal all the candy in Ninjago.

Gold helmet still shows Lloyd's true Ninja color

Golden katana

## Surprise achievement
Lloyd surprises everyone—especially himself—when he becomes the Golden Ninja, also known as the Ultimate Spinjitzu Master. Harnessing the power of the Golden Dragon, Lloyd saves Ninjago from the Overlord.

### DATA FILE
- **Also known as:** Master of Energy
- **Known for:** Leading Team Ninja when Wu is not around
- **Likes:** Comic books
- **Dislikes:** Meditating

# Lloyd

OVER THE YEARS, Lloyd has grown from pesky kid to Elemental Master to team leader. His relationship with his father, Garmadon, is complicated, but Lloyd always does the right thing. He is proud of how far he has come, and works hard every day to be the best Ninja and Master he can be.

## Uniting the Team

The Green Ninja combines all four elemental powers. In the same way, Lloyd unites the team. He can shoot energy bursts and charge batteries, but his greatest skill is encouraging his team.

### DID YOU KNOW?
Like his father and Wu, Lloyd has both Oni and Dragon blood, which means he can summon both Oni and Dragon forms.

## Race Car

Lloyd's green race car is built for speed. With its armor and built-in weapons, it's just as useful for a rescue mission as it is for winning a race!

Spoiler with golden blades

Race car transforms in three evolutions

# Dragons

**NINJAGO IS HOME** to many mythical creatures, none as ancient or powerful as dragons. When Ninja reach their True Potential, they can use their elemental powers to summon a unique dragon. Ninja who care for their dragons will be rewarded with the dragon's friendship and loyalty.

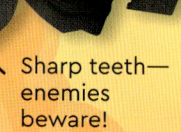

Sharp teeth—enemies beware!

Enormous wings enable super-fast flight

Surprise banner

Golden blades

# Legendary

Lloyd's Legendary Dragon is fierce but elegant as it soars through the air with ease. Its green scales and gold horns glimmer as Lloyd rides proudly on its back.

# Fire

Kai's Fire Dragon, like its master, evolves over time to gain new powers. The dragon has a scarily loud roar, but also likes to play.

Spiked tail is a deadly weapon

Courage banner

Fiery wings

Wings resemble water

# Water

Nya was always afraid of letting her team down. But when she accepts that failing is part of learning, she is finally able to summon her Water Dragon.

Wings of lightning

Speed banner

# Thunder

He might be the Master of Lightning, but Jay can harness the might of thunder, too. The Thunder Dragon fuses strength and speed— just like its Ninja master.

29

# How Do You Find Your Dragon Form?

ACCORDING TO THE ANCIENT Scroll of Quanish the Elder, "Only if the allies unite through a selfless act will main and might prevail the light, and dragon take flight." The Ninja and Wu find these instructions confusing, but that doesn't stop them from trying to figure them out!

## Paper Prophecy

Lloyd's mother, Misako, finds the Scroll of Quanish at the Library of Domu. It shows four Ninja poses, and contains a hidden message that is only revealed by folding over the four corners of the paper.

### DID YOU KNOW?
Wu and Garmadon knew Quanish when they were younger. They both say he was a fool.

## Four Ninja Moves

The ancient scroll shows four Ninja moves: jump up, kick back, whip around, and spin. The Ninja practice this sequence, but only during the final battle does it unlock their Dragon Form at last.

# Dragon Form

By performing the four moves, the Ninja summon their Golden Dragon wings and armor. The Golden Energy enhances their elemental powers.

*I'M LOVING THESE NEW WINGS, GUYS!*

Wings powered by elemental energy

*I GUESS QUANISH KNEW WHAT HE WAS TALKING ABOUT, AFTER ALL.*

Golden Dragon helmet

## Selfless Act

The Ninja choose to give up their elemental powers to protect the citizens of Ninjago. This selfless act unlocks the ultimate Dragon Form and unleashes the Golden Ultra Dragon.

31

# Myths and Legends

IN A LAND AS OLD as Ninjago, there are bound to be stories and tales about ancient monsters, curses, spirits, and realms. But no one really believes them, right?

## Wojira

Legend tells of an ancient storm spirit that used to rule the seas. Many believe that Wojira doesn't exist, but Nya can tell you it is very real indeed!

## Great Devourer

A giant serpent that will consume everything in existence? That can't be true! But the Great Devourer rises from its slumber at the Lost City of Ouroboros. Oh dear!

## Cursed Realm

Many tales speak of a desolate place where cursed souls dwell as ghosts. Lloyd once entered the Cursed Realm, and he is the only person to have ever come out alive.

## First Spinjitzu Master

There are many stories about the powerful First Spinjitzu Master—and most of them come from Wu. The First Spinjitzu Master was, in fact, his dad.

## Forbidden Scrolls

The two scrolls that detail a dark version of Spinjitzu have been lost for so long, some doubt their existence. But Zane knows they're real—he got rid of one of them!

## Ancient Weapons

These four legendary weapons are too powerful to be united by a single person. The Ninja protect the Golden Weapons—until Lord Garmadon is foolish enough to wield them all!

## Dark Island

Sailors and explorers have heard legends of a Dark Island formed from evil itself, but they've never found it. That is, until Lord Garmadon and the Overlord make it rise out of the sea.

## Traveler's Tea

Wu is a tea-lover, no doubt about it, but there is another sort of tea—Traveler's Tea—that can open portals and transport the tea-drinker to different places.

## The Oni

Also known as the Bringers of Doom, the Oni sound like they are from a scary story. But when they appear and start spreading Destruction, the Ninja realize how real they are.

*I'M IN CHARGE OF MY OWN DESTINY!*

## True Potential
Nya is hard on herself when she doesn't get things right. She finally unlocks her True Potential by accepting that failure is part of learning.

Nya controls mech from hidden cockpit

Samurai X Mech is a terrifying sight for enemies

## Samurai X
Nya has never been one to sit on the sidelines. Even before she became a Ninja, she was a force to be reckoned with! She created her own samurai armor and became the mysterious hero, Samurai X, saving the Ninja on more than one occasion!

### DATA FILE
- **Also known as:** Master of Water
- **Known for:** Being full of bright ideas
- **Likes:** Dancing
- **Dislikes:** Being called "Kai's sister"

# Nya

**NYA IS A FIERCE** warrior and clever inventor. Although she was a late addition to Team Ninja, the others have come to rely on her sensible advice, tech knowledge, and awesome battle skills. Her brother, Kai, and boyfriend, Jay, are constantly in awe of her incredible problem-solving skills.

## Water Spirit
Nya sacrifices her elemental powers and human form to save Ninjago. She defeats the storm spirit, Wojira, but ends up as part of the Endless Sea for a while.

### DID YOU KNOW?
Nya once died, but luckily her boyfriend, Jay, had a magical wish. He used it to undo the events that led to Nya's death.

## Elemental Master
After merging with the Endless Sea, Nya regains her human form—though it takes some time for her elemental powers to return. But soon enough, the Master of Water can once again control the rain and tides and summon the Water Dragon.

- Dragon responds to Nya's commands
- Strong tail can damage enemy vehicles

# Battle in the Deep

**EVIL PRINCE KALMAAR** has awoken the storm spirit, Wojira. The Ninja race deep beneath the waves to the Temple of the Endless Sea. The battle against Kalmaar's Maaray Guards is fierce, and Wojira is mightier than the Ninja feared. Will Nya be able to use the power of Water to stop Wojira?

Wave Amulet

Chains cannot stop Wojira

## Two Princes

The two princes of Merlopia find themselves on opposing sides of the battle. Kalmaar wants to overthrow all land-dwellers, while Benthomaar chooses to help the Ninja before it is too late.

Merlopian scientist Glutinous isn't used to being out of his lab

Kalmaar in the throne room

### DID YOU KNOW?
Wojira will only awaken when the ancient Storm Amulet and Wave Amulet are united and placed on her head.

STOP THE NINJA!

Maaray Guard

Prison building

Treasure room holds the Storm Amulet

I THINK I'LL HELP THE NINJA!

Prince Benthomaar

HUH?

Kai's mini-sub is no match for Wojira

**PRINCE KALMAAR'S WEAPON**
**1 TRIDENT**

37

"WHEREVER THERE IS LIGHT, THERE MUST ALWAYS BE SHADOW..."

## Golden Master

The Overlord uses the four Golden Weapons to forge the Golden Armor and become the legendary Golden Master. He uses his new Golden Power for a scary purpose.

## The Root of All Evil

The Ninja have come to realize that the Overlord was behind almost all the evil they've ever faced while defending Ninjago—and that's a lot of evil!

### DATA FILE

**Also known as:**
The Dark Lord

**Known for:**
Being pure evil

**Likes:**
Destruction and darkness

**Dislikes:**
Everything else

# Overlord

**ALL OF THE DARKNESS** in Ninjago can be traced back to the Overlord. A shadowy being, he wants to engulf Ninjago in darkness. The Ninja may battle the Overlord and his armies again and again, but he can never truly be defeated. He is as much a part of Ninjago as the earth and the clouds.

Green, glowing eyes

Unbreakable stone bodies

## Stone Army
The Overlord built the near-indestructible Stone Warriors from Dark Matter mined on the Island of Darkness.

### DID YOU KNOW?
The Overlord can take on various physical forms when he wants to, but nobody knows what he truly looks like.

Purple wings

Shadow-colored dragon hide

Sharp claws

## Power of Darkness
The Overlord's original form was a dragon, so it makes sense that he would take this form again. He attacks Ninjago as a dark dragon, breathing black fire.

39

# Enemies of Ninjago

**STOMPING, SLITHERING, OR FLOATING** their way through the streets, bands of villains are becoming a common sight in Ninjago. Whatever they're after, it's clear they are up to no good. The Ninja take it upon themselves to keep their world safe from these fearsome foes.

## Nindroids

The evil Overlord's robotic army is created based on Zane's blueprint. The Nindroids are faster and stronger than Zane, and have a cloaking feature to turn themselves invisible!

## Serpentine

An army of slithering snake tribes from underground, the Serpentine want to take control of Ninjago. Each tribe has its own scary snake power.

## Ghost Warriors

Banished spirits that escape from the Cursed Realm, the Ghost Warriors can possess people and objects.

# Sky Pirates

A motley crew of criminals, mercenaries, and oddballs, the Sky Pirates seek riches until they turn their attention to defeating the Ninja.

# Anacondrai

These villains worship the oldest Serpentine tribe, the Anacondrai. If they collect enough elemental powers they can perform a spell to transform themselves into true Anacondrai.

# Pyro Vipers

A long-gone Hypnobrai tribe brought back to life with the power of lava, the Pyro Vipers are out for revenge against whoever it was that defeated them last time.

# Dragon Hunters

In the ancient Realm of Oni and Dragons, this tribe hunts dragons to use their elemental powers. When they find the Ninja and learn about their powers, they can't believe their luck.

# Sons of Garmadon

This gang of misguided bikers worships Lord Garmadon! The group will do everything it can to install him as Emperor of Ninjago.

*YOU DON'T HAVE TO BE HUMAN TO BE A NINJA!*

Power source

## Cold as Ice
Zane's elemental powers grow a bit too frosty when he finds himself in the Never-Realm. Memories erased, Zane's serious personality turns seriously cold as the Nindroid becomes the fearsome Ice Emperor.

Titanium hair, too, of course!

### DID YOU KNOW?
Being a Nindroid doesn't stop Zane from falling in love. He is smitten with fellow android P.I.X.A.L. from the moment they meet!

Titanium muscles

## Titanium Zane
Zane's metallic body is actually his second body. His first was damaged when Zane sacrificed himself to save the team. Luckily, Zane was able to upload his consciousness into a new titanium body.

### DATA FILE
- **Also known as:** Master of Ice
- **Known for:** Vast intelligence and sarcasm
- **Likes:** P.I.X.A.L.
- **Dislikes:** His past as the Ice Emperor

# Zane

**THE NINJA OF ICE** is actually a Nindroid—part Ninja, part android. But that doesn't make Zane any less a part of the team. Zane's computer brain builds brilliant mission strategies, his mastery of technology can fix almost any problem, and his oddball personality makes his fellow Ninja love him even more.

## Nindroid Pal
As a Nindroid, Zane has many abilities his teammates find extremely useful. He can play music from his body, glow in the dark, and cool down food that is too hot!

Turbine engines

Jet shielded from Vengestone by P.I.X.A.L.'s shields

Extendable wings

## Golden Dragon Jet
Zane's Golden Dragon Jet is sleek and speedy. He pilots it into battle against members of the Crystal King's council, and even takes on some dangerous Dragonides in the skies above Ninjago.

# Fire Stone Mech Escape

**LOST IN A MAZE OF** underground tunnels with a Lava Monster nearby, Cole is close to giving up hope. That is, until he discovers a mech made of stone. Only a true Elemental Master can activate the mech, so Cole focuses with all his might. Can the mech help him find a way out?

## Solid as a Rock

Cole's plan relies on the mech's strength. If it can push the heavy mine carts up the track, they might just find their way to the surface.

GRAB THE BLADE!

I'LL GRAB YOU!

Enormous katana

Half stone, half fire

## DID YOU KNOW?
The Mech is powered by elemental energy, so Cole uses the power of Earth to bring it to life.

Kai sits in the mech's cockpit

BENEATH SHINTARO MOUNTAIN ARE **87** TUNNELS

Hand can blast rock

Ancient Shintaran stone

Powerful legs

HERE COMES THE MECH!

MECH? WHAT MECH? OH …

45

# Weapons

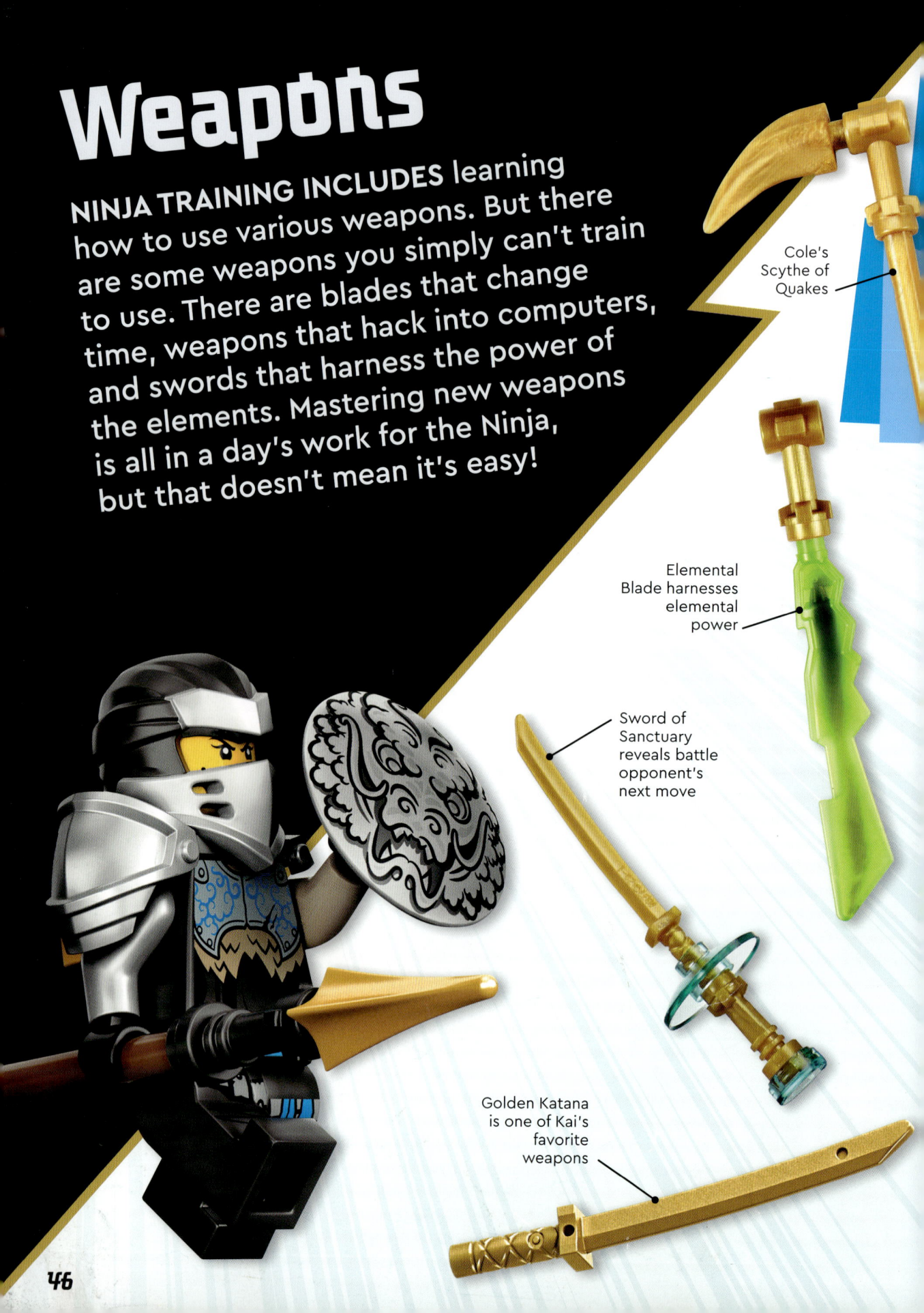

**NINJA TRAINING INCLUDES** learning how to use various weapons. But there are some weapons you simply can't train to use. There are blades that change time, weapons that hack into computers, and swords that harness the power of the elements. Mastering new weapons is all in a day's work for the Ninja, but that doesn't mean it's easy!

Cole's Scythe of Quakes

Elemental Blade harnesses elemental power

Sword of Sanctuary reveals battle opponent's next move

Golden Katana is one of Kai's favorite weapons

# Golden Weapons

These ancient weapons come as a set of four—but don't try to hold them all at the same time! On their own, they are powerful enough, channeling the four elements—Earth, Fire, Ice, and Lightning—but when united, they unleash the unthinkably powerful Tornado of Creation.

- Zane's Shurikens of Ice
- Kai's Sword of Fire
- Jay's Nunchucks of Lightning
- Techno Blade can interact with computers
- Time Blade can control time
- Dragon Blade summons the Fusion Dragon
- Shadow Blade is one of two Blades of Deliverance
- Lloyd's personal sword with ornamental tassel
- Staff of Dragons gives its wielder dragon abilities
- The Ivory Blade of Deliverance is paired with the Shadow Blade.
- Golden Spear is often Nya's weapon of choice

# Ghosted

Cole is a true team player. He once lost his own life while helping his friends at the Temple of Spinjitzu. Cole became a ghost for a while, though he regained his human form in the end.

*LET'S ROCK AND ROLL!*

Cool guitar shirt

# Music Lover

Cole was born into a family of performers, but he left to forge his own destiny. Still, on his days off, he enjoys listening to music (especially soft rock), singing, and even dancing with his Ninja friends.

## DATA FILE

- **Also known as:** Master of Earth
- **Known for:** His loyalty and poor cooking skills
- **Likes:** Cake
- **Dislikes:** Letting his friends down

# Cole

**STRONG AND DEPENDABLE** as the earth itself, Cole embodies his elemental power. When Team Ninja disperses, Cole never gives up hope that they will get back together. Cole takes charge and rebuilds the team, leading it to attack a mysterious and powerful villain, the Crystal King.

Cole is rarely seen without his black gloves

## Down to Earth
As the Master of Earth, Cole has many mighty powers. He can manipulate the ground itself, summoning earthquakes or creating cracks in the rock. He can also Earth Punch his way through almost anything.

### DID YOU KNOW?
As the Ninja of Earth, Cole wears a lot of black. His favorite color, however, is orange. His Ninja suits often feature orange details.

Lightning blades in Dragon Attack mode

## Dragon Cruiser
Cole's vehicles are among the most powerful of all Ninja vehicles. His Dragon Cruiser has two lightning blades, which come in very handy during the battle against the Crystal King's Vengestone Army.

Ultra-durable wheels

49

# Ultra Sonic Raider

ONE VEHICLE IN FOUR parts, the Ultra Sonic Raider is superfast and fitted with weapons galore. The Ninja rely on the Raider to battle the Great Devourer. Its sonic cannons help weaken the evil serpent.

Raider Jet piloted by Jay

Cole's side bike detaches from the main vehicle

Tank treads for all terrain

A hidden tire slips into place when the main vehicle splits off

## Raider Jet

The top section of the Ultra Sonic Raider can lift off and fly! The Raider Jet's twin propellers help it fly at lightning speeds.

Blades help steer while in flight mode

Golden blades lift to form jet's wings

Sonic cannon

### Racing Raider
The Ninja ride the Ultra Sonic Raider to victory in the Ninjaball Run, Ninjago's fiercest race.

### DID YOU KNOW?
The Raider Jet survived epic battles and a dangerous race, only to fall apart when a butterfly landed on its wings.

51

# Mastering the Elements

**GENERATIONS OF ELEMENTAL MASTERS** have passed down their powers over the centuries, using their mystical abilities to protect Ninjago.

### DID YOU KNOW?
Fire, Lightning, Ice, and Earth are the four fundamental Elements of Creation.

### Master of Fire
Kai harnesses the fierce, unpredictable element of Fire. He loves the heat of battle.

### Master of Energy
Lloyd can create energy beam blasts and force fields. He can even use bursts of energy to fly!

### Master of Ice
Zane's powers allow him to freeze his enemies in their tracks—and then whip up a perfect ice crea[m]

### Master of Water
Nya is adaptable, like water. She uses water to make shields, weapons, and even heavy rain.

### Master of Lightning
Harnessing the power of electricity is highly useful—especially when you're a tech genius like Jay!

### Master of Earth
Cole can blast any rock out of his way. He can also generate earthquakes and build mountain[s]

52

**Master of Shadow**
Shade blends perfectly into the shadows. He can also travel between any two shadows.

**Master of Smoke**
Ash can transform into swirls of wispy smoke, which makes him very hard to pin down.

**Master of Nature**
Bolobu can grow any plant, anywhere—and he can command trees and vines during battle.

**Master of Amber**
Skylor can absorb other people's elemental powers. She can even combine two or more together!

**Master of Gravity**
Facing Gravis in battle can be a strange experience. Which way is up? Only Gravis knows.

**Master of Speed**
Griffin Turner is fast in every way: he thinks fast, acts fast, moves fast, and even talks fast!

**Master of Sound**
Jacob Pevsner is blind, but his sense of sound is so powerful it can create sonic weapons.

**Master of Metal**
Not only is Karlof's body extra durable, he can even turn it into hard metal!

**Master of Mind**
Neuro can read and manipulate minds. But knowing everyone's secrets isn't always fun.

# Who Will Win the Tournament of Elements?

**THE NINJA RECEIVE** a mysterious invitation to take part in the Tournament of Elements on a remote island. They face off against other Elemental Masters and compete in thrilling challenges. But as in most adventures in Ninjago, things are not all as they seem …

**TOURNAMENT OF ELEMENTS — 16 CONTESTANTS**

Temple houses several Jadeblades

Spinning obstacle

## Round One

The first round of the tournament is a hunt for hidden Jadeblades. There are enough Jadeblades hidden on the island for every contestant—apart from one. The race is on!

## DID YOU KNOW?

No one was ever supposed to win the tournament. Master Chen set the whole thing up as a way to steal elemental powers.

# Good and Evil

MANY FORCES OF DARKNESS threaten Ninjago, but sometimes the Ninja must confront a danger much closer to home: the evil within themselves.

## Good Wu

Wu is a wise and powerful mentor. The Ninja turn to him for advice and answers, so when he is turned into an evil cyborg version of himself the Ninja know they are in serious trouble!

*YOU CAN'T OUT-NINJA ME!*

## Techno Wu

Having Wu as an enemy is about as bad as it can get. He knows the Ninja's strengths, weaknesses, techniques, and secrets. Worst of all, the Ninja don't want to battle against their Master!

## Master Garmadon

Garmadon as a good guy takes some getting used to, especially for Lloyd. But the Ninja come to value their new ally. Master Garmadon has a lot of wisdom to share.

*I LOOK GOOD IN WHITE, DON'T I?*

## Lord Garmadon

Garmadon is one of Ninjago's fiercest foes. He was infected with evil as a child and has a thirst for power, but deep down he has a softer side, especially for Lloyd.

*NINJA RULE, GHOSTS DROOL!*

**DID YOU KNOW?**
Morro has always been obsessed with the idea of becoming the Green Ninja. That's why he's so set on defeating Lloyd.

## Lloyd

It took a lot of courage for Lloyd to banish his father to the Cursed Realm. But the need for courage didn't end there. Lloyd accidentally unleashed Morro at the same time!

## Morro

Morro was Wu's first ever pupil, who went bad. He possessed Lloyd, making him attack the other Ninja. Lloyd was relieved when he was finally freed.

## Zane

Zane is known for being the cool, calm, and collected Ninja. He's worked hard to develop a sense of humor over the years, but there's no denying it—he's the serious one.

*I'M AS COLD AS ICE.*

## Ice Emperor

As the Ice Emperor, Zane's even more frosty than usual. In fact, he's frozen solid. Without his memories, Ice Emperor Zane is far from his kind-hearted self.

# Ice Dragon Attack

**A COLD WAR RAGES** across the frozen wastelands of the Never-Realm. The merciless Ice Emperor sends out his dragon to freeze nearby villages. But Lloyd has discovered a secret: the Ice Emperor is none other than his friend Zane! And it's time to help the Ninja of Ice remember who he truly is.

## DID YOU KNOW?
Of all the realms of Ninjago, the Never-Realm is the most remote—and the only one Wu's father warned him to avoid.

Flag of the Ice Emperor

### General Vex
This power-hungry warrior leads the army of Blizzard Samurai, but he is also secretly manipulating the Ice Emperor to do his bidding.

### Blizzard Sword Master
A ferocious soldier of the Ice Emperor's army, the Blizzard Sword Master is definitely cold as ice, seeing as he is made from snow and ice.

"ICE, ICE, BABY!"

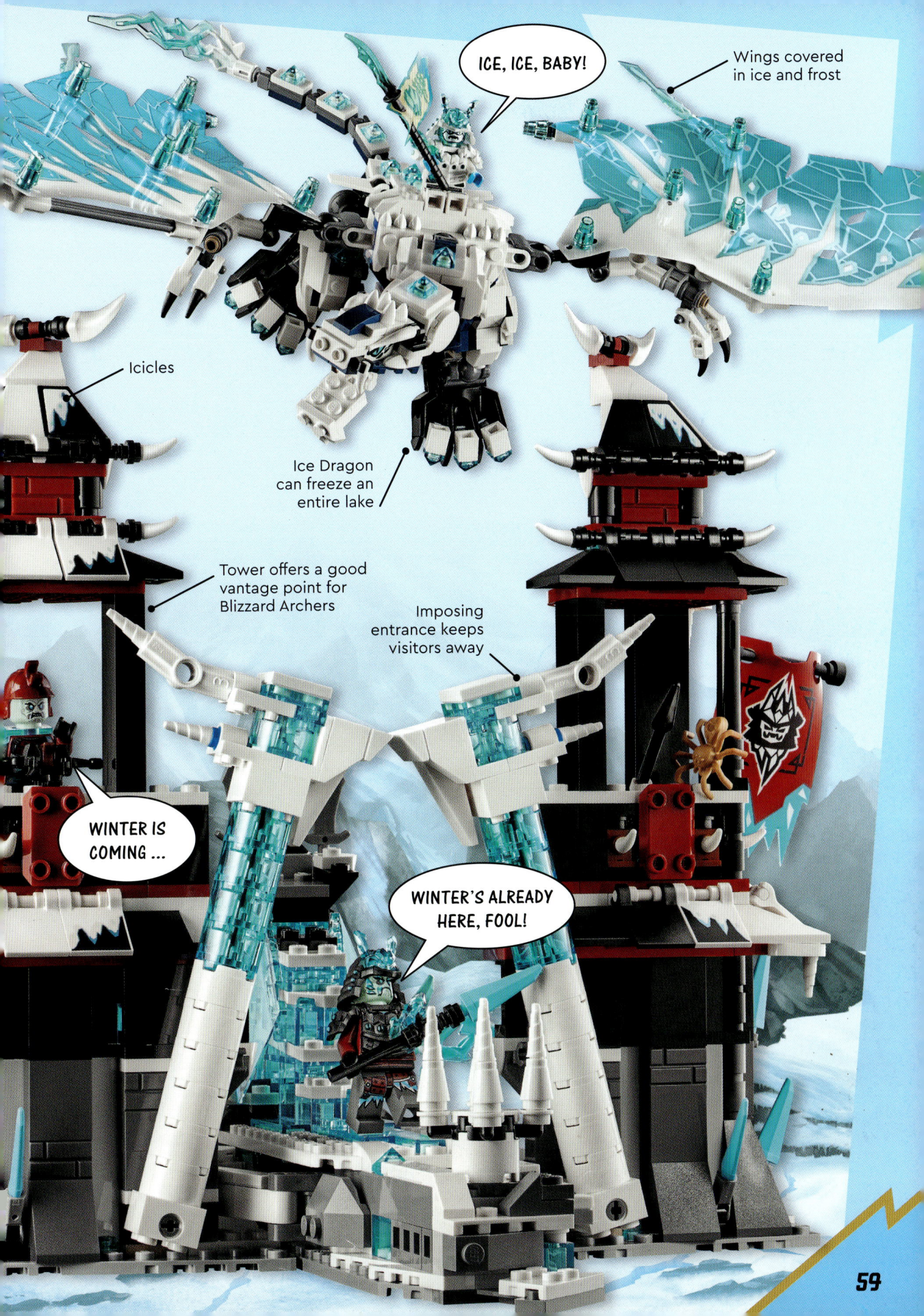

Wings covered in ice and frost

Icicles

Ice Dragon can freeze an entire lake

Tower offers a good vantage point for Blizzard Archers

Imposing entrance keeps visitors away

"WINTER IS COMING ..."

"WINTER'S ALREADY HERE, FOOL!"

59

*IMITATION IS THE HIGHEST FORM OF FLATTERY!*

## Controlling Colossus

Skylor risks her life to save Ninjago from the stone giant, Colossus. She absorbs Lord Garmadon's powers and uses them to control Colossus, though the strength of the powers ends up poisoning her.

## Using Her Noodle

Skylor inherited the family noodle business from her evil father, Master Chen. She now uses it to do good—from feeding hungry people to boosting morale during hard times.

### DATA FILE

- **Also known as:** Master of Amber
- **Known for:** Her noodle dishes
- **Likes:** Being part of a team
- **Dislikes:** When people don't trust her

# Skylor

**SKYLOR HAS A RARE** ability—she can absorb and use other elemental powers through touch. She is quick to join her Ninja friends in battle, though sometimes the powers she absorbs can overwhelm her! When she's not aiding the Ninja, Skylor can be found cooking up delicious noodle recipes.

### DID YOU KNOW?
Skylor is proud of her restaurant, the Noodle House. One of the most popular items on the menu is the Puffy Potsticker.

Dark katana

Skylor has excellent aim with a shuriken

## Former Spy
Skylor was a spy for her father, Master Chen, at the Tournament of Elements, where she first met the Ninja. But when she realized her dad was the bad guy, Skylor did everything she could to help the Ninja and stop her father.

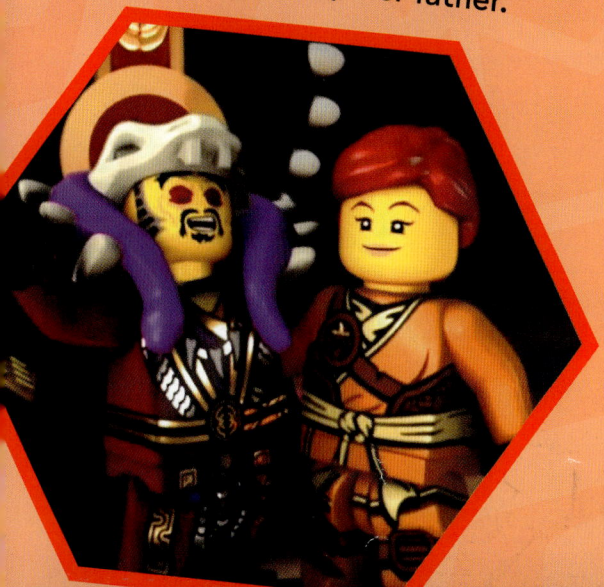

## Fierce Fighter
Skylor replicates the elemental powers of others, but it doesn't mean she has them forever. After a while, they get used up. So Skylor also relies on her own skills and training during battle.

# Ninja Gamers

**VIDEO GAMES ARE PART** of life, or so the Ninja keep trying to convince Master Wu. One day, a new game, *Prime Empire*, soars in popularity across Ninjago—everyone seems to be playing it. Until one by one the gamers start to disappear!

**INSIDE PRIME EMPIRE 4 LIVES**

Arcade games are popular in Ninjago

Lloyd's avatar has a unique look

Display shows how many lives Lloyd has left

Lloyd in Digi Robes, ready to enter the game

## Prime Empire

The idea of going inside a video game would appeal to any gaming fan! But when *Prime Empire* starts sucking players into the game, the Ninja quickly discover that the dangers they are facing are completely real.

# Amazing Avatars

Avatar versions of the Ninja and their friends populate the world of *Prime Empire*. Some avatars are pretty similar to their real-life counterparts while others look totally different!

Cole loves his avatar's bushy mustache!

# Digi Robes

Inside *Prime Empire*, the Ninja earn credits to upgrade their weapons and outfits. With enough credits they can get advanced Digi Robes with useful energy signal displays.

Jay's avatar is a legend within the game

## Dojo Games Room

In their free time, the Ninja like to hang out and play video games. It's not as stressful as the mission inside *Prime Empire*—unless Jay starts getting too competitive.

Red Visors carry red katanas

# Red Visors

Unfortunately, inside *Prime Empire* the Ninja can't just play games and dress up their avatars in silly costumes. An army of Red Visors are after them under orders from their angry leader, Unagami.

63

"STEP INTO MY WORKSHOP"

## Under the Overlord

Just after P.I.X.A.L. first met the Ninja, the Digital Overlord took control of her systems. P.I.X.A.L. was a formidable enemy—the Ninja were very relieved to get her back on their side!

## Not Quite Human

As a Nindroid, P.I.X.A.L. wasn't programmed to feel emotions, or to care about things like loyalty or compassion. But her strong friendship with Zane has helped her develop a more human personality.

### DATA FILE

- **Also known as:** Primary Interactive X-ternal Assistant Lifeform
- **Known for:** Her brilliant mind
- **Likes:** Zane
- **Dislikes:** Violence

# P.I.X.A.L.

A NINDROID CREATED to be a personal assistant, P.I.X.A.L. has evolved into so much more! Not only does she charge into battle as the new Samurai X, but she also uses her considerable brainpower to come up with inventions, battle plans, high-tech gadgets, and vehicles for the Ninja.

## Workshop

P.I.X.A.L. likes to tinker with machines, and being a machine herself she is pretty good at it! Her workshop is equipped with top-of-the-range computers and tools designed by P.I.X.A.L. herself.

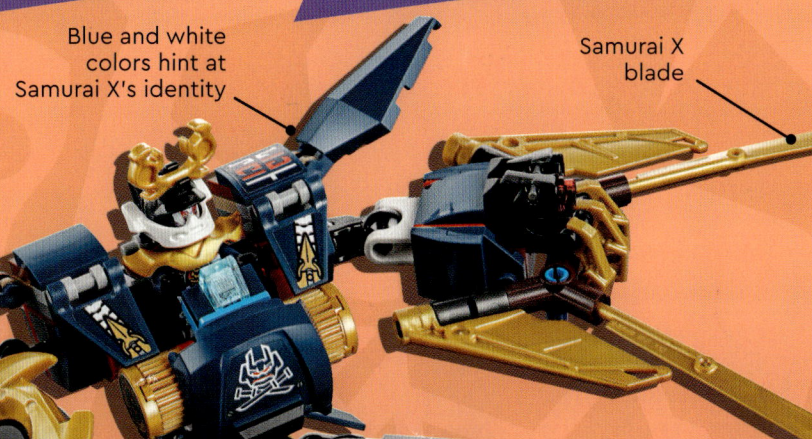

Blue and white colors hint at Samurai X's identity

Samurai X blade

### DID YOU KNOW?

At one point, P.I.X.A.L. only existed inside Zane's head! When she had no body, Zane loaded her hard drive inside himself.

## New Samurai X

When P.I.X.A.L. decides to join the Ninja in battle, she uploads her artificial intelligence into the Samurai X mech. In keeping with Samurai X tradition, P.I.X.A.L. keeps her identity secret for a while.

Large feet for balance

# How Do Ninja Train?

**NINJA TRAINING SESSIONS 4 A DAY**

IT TAKES HARD WORK and determination to be a Ninja. With skills to learn and powers to master, they spend hours and hours at the training dojo. But it doesn't stop there. A true Ninja will use every life experience to learn something new.

Target practice

Punchbag

Snake pit to practice leaping and jumping

Training mech

## Suit Up!

Different skills require different gear. Battling with weapons? Grab your Kendo armor. Practicing your Elemental Ice Powers? Don't forget your gloves.

### Spinjitzu Suit
Spinjitzu training is tough. Choose a durable suit that can contain the fiery power of Spinjitzu.

### Kendo Armor
Protective Kendo armor is very useful when training in weapons combat.

### Training Robes
For regular training sessions, this soft, simple suit is perfect. It has no armor or accessories to weigh you down.

Training slide teaches you how to make a stylish entrance

# Practice

There's no denying that practice makes perfect. Wu sets up the training arena with all sorts of challenges and traps. The Ninja don't stop until they get it right.

Practice Spinjitzu in an open area

# Take Your Time

A Ninja understands that mastering a new skill takes time. Wu teaches his students the value of patience. Sometimes it's not all about action.

# Have Fun!

It won't feel like hard work if you're having a good time. The Ninja are friends first and foremost, so every training session feels like hanging out with your pals.

*A GIRL'S GOT TO HAVE HER SECRETS!*

Vengestone crystal gives Harumi super-strength

## True Colors
Although she keeps teaming up with the bad guys, Harumi proves she's not evil to the core when she sacrifices herself to save an innocent family.

### DID YOU KNOW?
Harumi was adopted into the royal family. Her biological parents were killed by the Great Devourer when Harumi was a child.

Royal attire

## Jade Princess
When your parents are the Emperor and Empress of Ninjago, it's hard to escape the spotlight. Harumi sometimes wipes off her traditional makeup, changes her clothes, and sneaks out of the palace grounds.

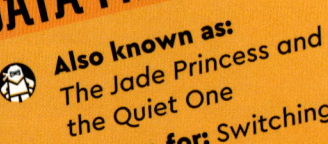

### DATA FILE
- **Also known as:** The Jade Princess and the Quiet One
- **Known for:** Switching sides
- **Likes:** Scheming
- **Dislikes:** Lloyd and the Ninja

# Princess Harumi

SHE MIGHT BE Ninjago royalty, but Harumi isn't your average princess. For one thing, she's against the Ninja and blames them for her parents' deaths. Harumi works hard to gain the Ninja's trust while she embarks on a dark and dangerous mission to seek revenge.

## Bad Friends

Harumi idolized Lord Garmadon for defeating the Great Devourer. Her devotion to him was so strong, she helped bring him back from the Cursed Realm to lead the Sons of Garmadon.

## Master of Disguise

With her excellent acting skills and convincing disguises, Harumi has fooled the Ninja on many occasions. Lloyd is shocked to discover that it was Harumi behind the Kabuki Mask all along!

## Test of Friendship

Harumi's friendship with Lloyd has had its ups and downs. But there's always been a connection there. When Harumi finally realizes the Ninja are not her enemy, Lloyd welcomes her with open arms.

# Can the Ninja Stop the Sons of Garmadon?

**IF THE NINJA BELIEVE** Lord Garmadon is truly gone, they need to think again. A gang that worships Garmadon is plotting to bring the villain back from the Departed Realm. The Ninja fight with all their power, believing that no good will come if Garmadon returns.

Roof transforms into a different shape

Rebuilt Temple of Resurrection

CAN YOU RESURRECT ME NEXT?

## Faithful Followers

The Sons of Garmadon are members of an angry biker gang who believe Garmadon will be the savior of Ninjago. When they are not resurrecting evil lords and battling Ninja, they enjoy karaoke and organizing street races.

Prisoner who never escaped

WHERE DID ALL THESE ARMS COME FROM?

## DID YOU KNOW?
Uniting the three Oni Masks will open a portal to the Departed Realm and bring back someone with Oni blood.

Oni masks housed inside

Spider web conceals prison chamber

Fanged doorway

Moat filled with piranhas

No one knows who this baby is

71

# Teamwork

**NO MATTER HOW** powerful the Ninja are, they won't get anywhere if they can't work as a team. Each Ninja has their own strengths and weaknesses. Working together unites their skills and makes them an unstoppable force.

Cole's Earth Car gives the mech drill arms

Kai's Fire Mech gets supersized when the Ninja team up

## Mech Team-Up

What do you get when you cross a Fire Mech, Lightning Jet, Earth Car, and Ultra Sonic Raider? The Ninja can't wait to find out. They are thrilled with their Ultra Combo Mech, which has four powers—and lots of weapons!

ULTRA COMBO MECH FITS 4 NINJA

*THAT'S A BIG MECH ...*

Mech seems a little small now

### Mech Match
The Cobra Mechanic prides himself on his battle machine creations. But once he spots the Ultra Combo Mech, he doesn't think his mech is that impressive any more!

*BEEP. BOOP. TEAMWORK. BEEEEP.*

Wu bot

Jay's Lightning Jet forms the top of the mech

Four golden heads

Harnesses Golden Power

Zane's Ultra Sonic Raider splits to form the legs

Boa Destructor is not used to being defeated

GO, TEAM NINJA!

I THINK I'VE MET MY MATCH!

# Ninja Unite

The Ninja discover that the only way to save Ninjago from the Crystal King is to work as a team. As one, they release their elemental powers using the Golden Weapons. This act of teamwork unleashes the magnificent Golden Ultra Dragon.

## DID YOU KNOW?

In the final battle against the Crystal King, the Ninja are busy with Ninja duties, so it's P.I.X.A.L. who pilots the Combo Mech!

73

# Battle at Sea!

**A STORMY BATTLE RAGES** when Wu goes missing. The Ninja set out to sea in search of him, but are found by a mysterious tribe known as the Keepers. The Keepers want to protect the precious Storm Amulet but the Ninja just want to rescue Wu!

## DID YOU KNOW?
The Keepers are an ancient tribe that lives on an uncharted island. They have storm powers and can control lightning.

- Sail provides balance on the stormy sea
- Rumble Keeper holds tribal shield
- Each section has its own sail
- Spiked blades for damaging enemy boats
- Tribal markings
- Prison boat splits off from main section

*"WHY ARE WE BATTLING THE NINJA AGAIN?"*

*"THEY HAVE SEEN THE AMULET!"*

*"SO WHY ARE THEY CHASING US?"*

# Speedy at Sea

Kai steers the Ninja catamaran through the choppy waves. A sleek sail helps the nimble boat balance, while its twin propellers gather speed. Kai can draw the sides of the catamaran closer to squeeze through narrow spaces.

*Fire shooter*

*Flame design on sail*

*Strong hull withstands submerged rocks*

*Catamaran can transform into a speedboat*

AMULET? WHAT AMULET?

**CAPTURED BY THE KEEPERS**
**9**
**PEOPLE**

# Captured?

The Keepers' prison boat is fitted with blades to keep captives captive. But the Ninja always have a trick (or 10) up their sleeves, so you can be sure they won't stay trapped for long!

75

# Armies Attack!

**EVERY SUCCESSFUL VILLAIN** has a faithful army at their command. And what the foot soldiers lack in brainpower they make up for in muscle! The invading armies might wield powerful weapons, but if the Ninja get their way these sinister soldiers won't be around for long.

*NINJAGO INVASIONS ?! TOO MANY TO COUNT*

*WE'VE GOT A BONE TO PICK WITH YOU!*

## Skulkin

These bony baddies are from the underworld. They follow Lord Garmadon on his quest to unite the Golden Weapons. Being dead already gives them a huge advantage in battle!

## Vermillion Warriors

The children of the Great Devourer follow in their parent's tail-tracks. Formed from slithering snakes, these soldiers won't sssstop until they've consumed everything!

*NINJAGO IS HISSSSSTORY ...*

### DID YOU KNOW?
While the Skulkin and the Awakened Warriors look pretty similar, they do not appreciate being mistaken for each other!

# Awakened Warriors

The Skull Sorcerer's army is made up of dead warriors that are brought back to life. Their green eyes glow in the darkness of the Dungeons of Shintaro, giving the Ninja a fright!

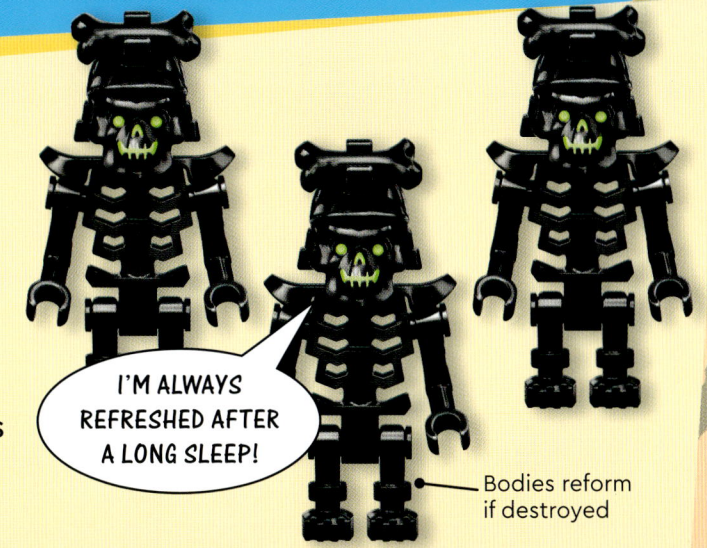

*I'M ALWAYS REFRESHED AFTER A LONG SLEEP!*

Bodies reform if destroyed

# Blizzard Samurai

These ice-cold warriors serve the Ice Emperor. The army is made up of warriors, sword masters, and archers, and led by the scary General Vex.

Blizzard Warriors have no sense of humor

# Maaray Guards

*WATER YOU GOING TO DO ABOUT ME, NINJA?*

An army of eel-like soldiers, the Maaray Guards follow the orders of King Kalmaar to defend their kingdom, Merlopia. The guards are strong swimmers and skilled with their trident and blade weapons.

Official guard uniform of Merlopia

77

# Escape from the Dungeons of Shintaro

**THE NINJA ARE TRAPPED** in the lava-filled Dungeons of Shintaro. The Skull Sorcerer and his creepy army of Awakened Warriors have no plans to let them go. Cole is the Ninja's only hope. Can he stop the Skull Sorcerer in time?

**IMPRISONED IN THE DUNGEONS**
**5**
**NINJA**

Princess Vania of Shintaro

Ginkle of underground Geckle tribe

The Dungeons lie far below ground

Rock path over the lava pit

## Scary Skull

The Skull Sorcerer gets his powers from the ancient Skull of Hazza D'ur, an evil sorcerer from centuries ago. The skull glows with dark magic and enables the Skull Sorcerer to control an army of undead skeletons—and to summon the terrible dragon, Grief-Bringer.

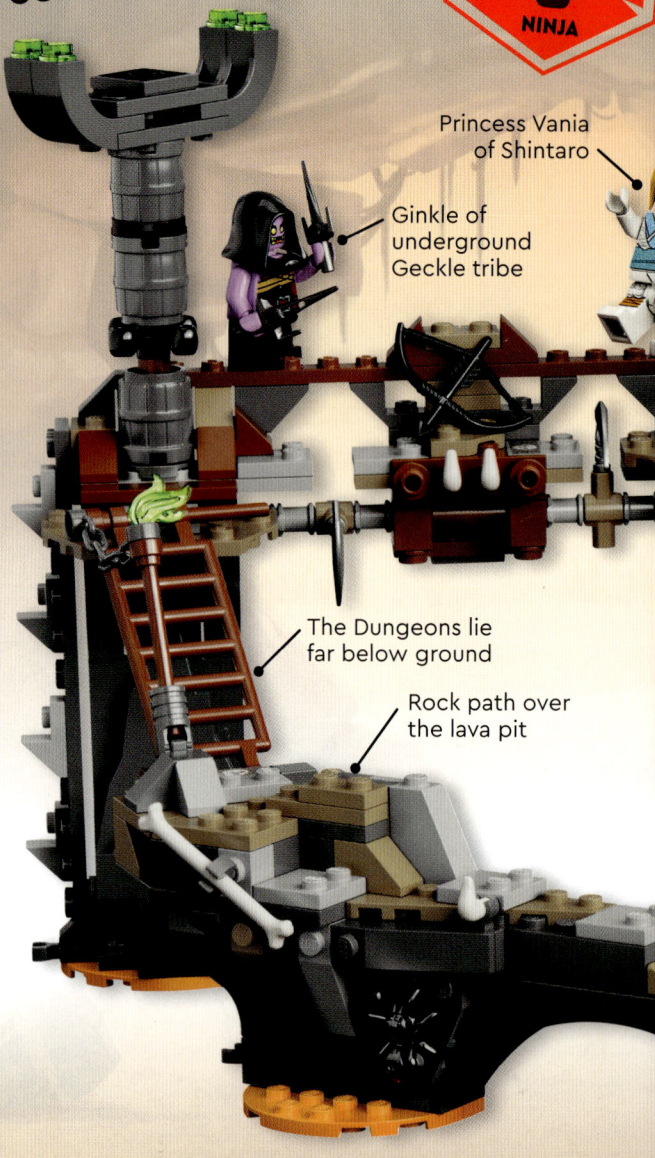

The Skull Keep

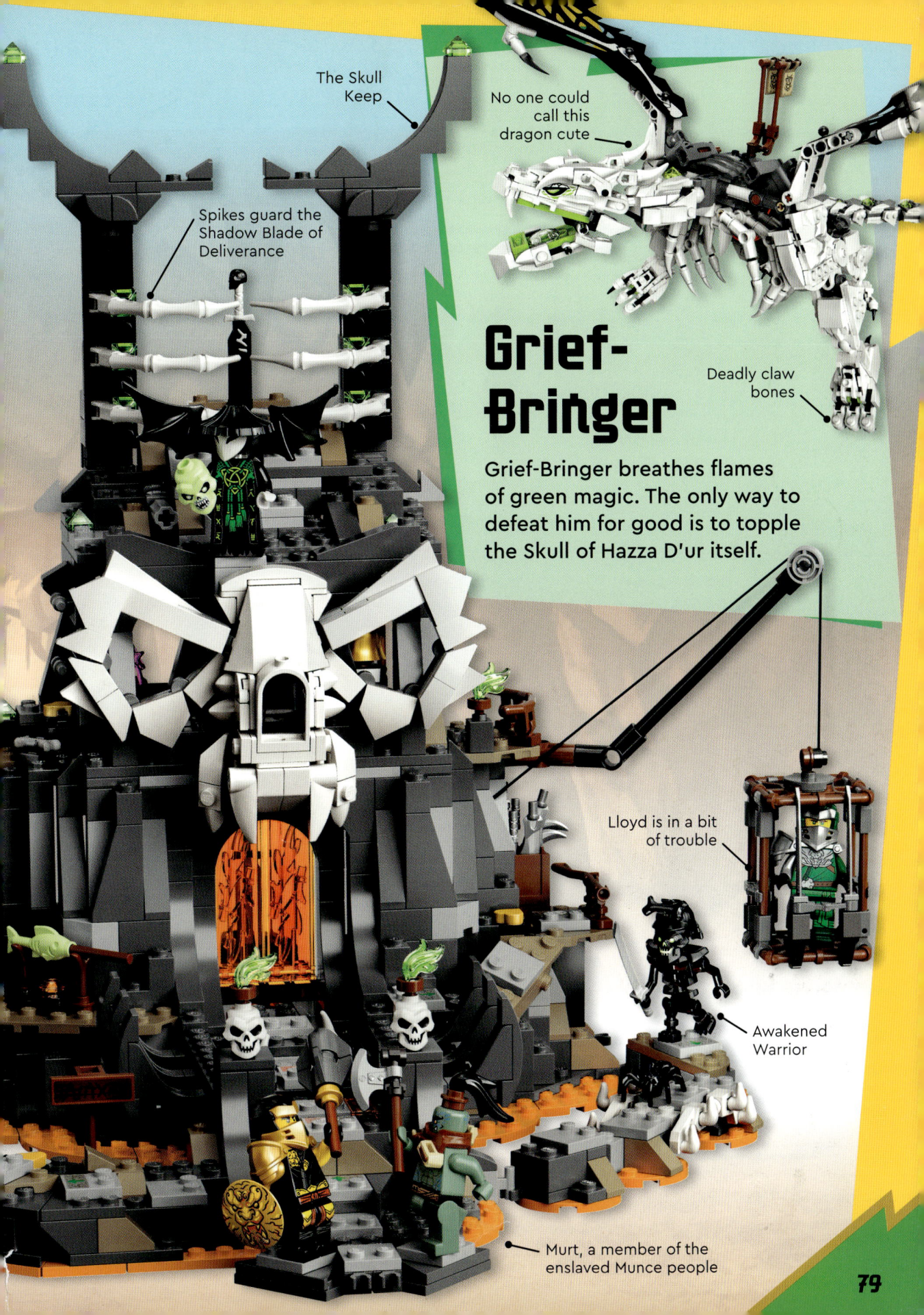

No one could call this dragon cute

Spikes guard the Shadow Blade of Deliverance

# Grief-Bringer

Deadly claw bones

Grief-Bringer breathes flames of green magic. The only way to defeat him for good is to topple the Skull of Hazza D'ur itself.

Lloyd is in a bit of trouble

Awakened Warrior

Murt, a member of the enslaved Munce people

# What Are the Fundamental Ninja Virtues?

**WU FINDS IT IRRITATING** when his Ninja students don't remember his lessons. Sometimes they'd rather eat cookies and crack bricks than meditate on Ninja values. But when the Ninja are in the middle of an adventure, they must recall Wu's teachings to overcome challenges.

## Curiosity

A good Ninja asks questions about the things around them. When Nya hears a strange noise, her curiosity helps the team discover that Wu is in danger!

## Balance

The balance of good and evil is always shifting in Ninjago, but there's another kind of balance that Cole calls upon to cross a pit of lava. He clears his mind, which balances his body so he can pass safely.

# Wisdom

Master Wu isn't the only wise Ninja. Lloyd calls on his own brain to help during a wild race, when he realizes that the key to winning is thinking smart—not driving fast!

# Honesty

Honesty is the best policy, but telling a white lie can be tempting. Luckily, Jay chooses to tell the truth to the Ice Emperor—and his honesty wins the Ninja their freedom.

# Generosity

A Ninja helps others, even if it comes at a cost. Zane gives his own arm to repair a catapult during a battle, and ensures everyone is safe before the Ninja depart.

# Courage

Courage might be a basic Ninja virtue, but it's not always easy to summon it. Especially when facing four huge dragons! Kai bravely steps up to battle the dragons and rescue Wu.

# Monastery of Spinjitzu

THE MONASTERY OF Spinjitzu was the home of the First Spinjitzu Master and his two sons, Wu and Garmadon. Now it is where the Ninja live, train, and play video games. The ancient building sits right at the top of the Mountains of Impossible Height, which everyone agrees is a great location—apart from the mail carrier, that is!

## History Lesson

The Ninjago history mural shows important events from Ninjago's history. It reminds the Ninja what they are training for.

## Salad Skills

One of Wu's training methods is tasking the Ninja to chop fruit. This hones their knife skills—and also teaches them how to make a good fruit salad!

Strong locking mechanism on gate to keep enemies out

## DID YOU KNOW?
The Monastery has been burned down once, rebuilt, and then demolished again. Kai once set the ruins on fire as well. Oops!

Traditional shrine architecture

Banners of Spinjitzu

## Weapons Wall
The four Golden Weapons belong in the Monastery. And of course, each one is fully guarded! The Sword of Fire is hidden behind a fake wall and pretty potted plant.

Lately, mysterious things have been happening at the Monastery

Black shingle roof

TEA, ANYONE?

A flying chicken trap protects the Golden Weapons

CAN YOU BEAT ME, ZANE?

SHUR-I-KEN!

83

## Crystal Council

The council of the Crystal King is made up of some of the Ninja's worst enemies. Led by Harumi, the council includes Vangelis, Aspheera, Pythor, Mr. F, and the Mechanic.

**CRYSTAL POWER!**

Crystal energy

Vengestone

## Vengestone Army

The Crystal King has built a fearsome army of stone warriors. When he brings them to life, they are powered up by the Vengestone that runs through them.

### DATA FILE

- **Also known as:** A mysterious villain
- **Known for:** Scary, horned appearance
- **Likes:** Vengestone crystal
- **Dislikes:** When his orders aren't followed

# Crystal King

**A STRANGE NEW ENEMY** surfaces with just one aim—the total destruction of Ninjago. The Crystal King has been plotting from the shadows to gather a following of Ninjago's greatest villains. He has secretly built himself a powerful army. There's only one thing standing in his way: a small team of Ninja warriors!

## Crystal Infection

The Vengestone Warriors spread a crystal infection across Ninjago. Each of the infected become another warrior in the Crystal King's army.

### DID YOU KNOW?

Vengestone is a crystal that cancels elemental powers. The Crystal King channels his dark energy through a large crystal in his palace.

## Corrupted Weapons

The four Golden Weapons were used to create Ninjago. But the Crystal King corrupts their power and transforms them into the Weapons of Destruction.

Scythe of Quakes

Crystal energy transforms the weapons

85

# The Final Battle

AN ANCIENT PROPHECY tells of the Final Battle, where a single moment will decide the fate of Ninjago. And now darkness is coming, consuming everything, which can mean only one thing—that moment has arrived! But the Ninja are terribly outnumbered ...

**IN THE FINAL BATTLE THERE ARE JUST 2 SIDES**

## Unlikely Allies

As the Crystal King and his Vengestone Army advance on Ninjago, the Ninja are joined by former allies and—to their great surprise—old enemies, too! United they stand, ready to defend their world from evil.

## Misako

Lloyd's mother has been a longtime ally of the Ninja, but now she takes to the battlefield to help.

# Heroes Unite

Skylor and former video game character Okino join the battle. They've never met, but there's no time for introductions as they take on a Vengestone Warrior.

## Ronin

Thief Ronin hasn't always been on the side of good, but he comes through in the end, joining the Ninja in the Final Battle.

## Vania

The Princess of Shintaro rushes to join the Ninja. She brings along the army of Shintaro and her dragon, Chompy.

## Dareth

The Ninja haven't seen their old pal the Brown Ninja for a while, but Dareth leaps fearlessly into battle!

## Ultra Violet

She may have been a general in the Sons of Garmadon, but Ultra Violet chooses the right side now, charging into battle against the Vengestone Army.

# Attack on the Crystal Palace

**THE CRYSTAL KING'S** plan to crystallize the whole of Ninjago is almost complete. His floating palace seems impenetrable. Can the Ninja overcome the weakening powers of Vengestone crystal and defeat the Crystal King before it's too late?

Dragonide with crystal powers

Nunchucks of Lightning

## Scary Centaur

All hope seems lost when the Crystal King takes the form of a huge centaur—with four legs and crystal wings. But a Ninja never gives up. Let's hope that's enough!

Rock body embedded with crystals

Fists blast crystal energy

The palace is full of traps

Throne Room of the Crystal King

Floating prison

## DID YOU KNOW?
The central crystal in the Throne Room controls the entire Crystal Army. If it shatters, the army will instantly power down.

Blade of Fire

Vengestone crystal

### Golden Threat
It looks like nothing can stop the Crystal King. But when the Ninja unite the Golden Weapons, elemental energies combine to create the Golden Ultra Dragon. For the first time, the Crystal King shows a hint of fear.

YOUR POWERS DON'T WORK AROUND HERE!

AT LEAST MY BRAIN DOES!

89

# A New Reality

**LIFE IN NINJAGO** never stays the same for long, and now there are new realms appearing and disappearing at random! This new reality means that everything feels a little uncertain. And as for the Ninja … what Ninja?

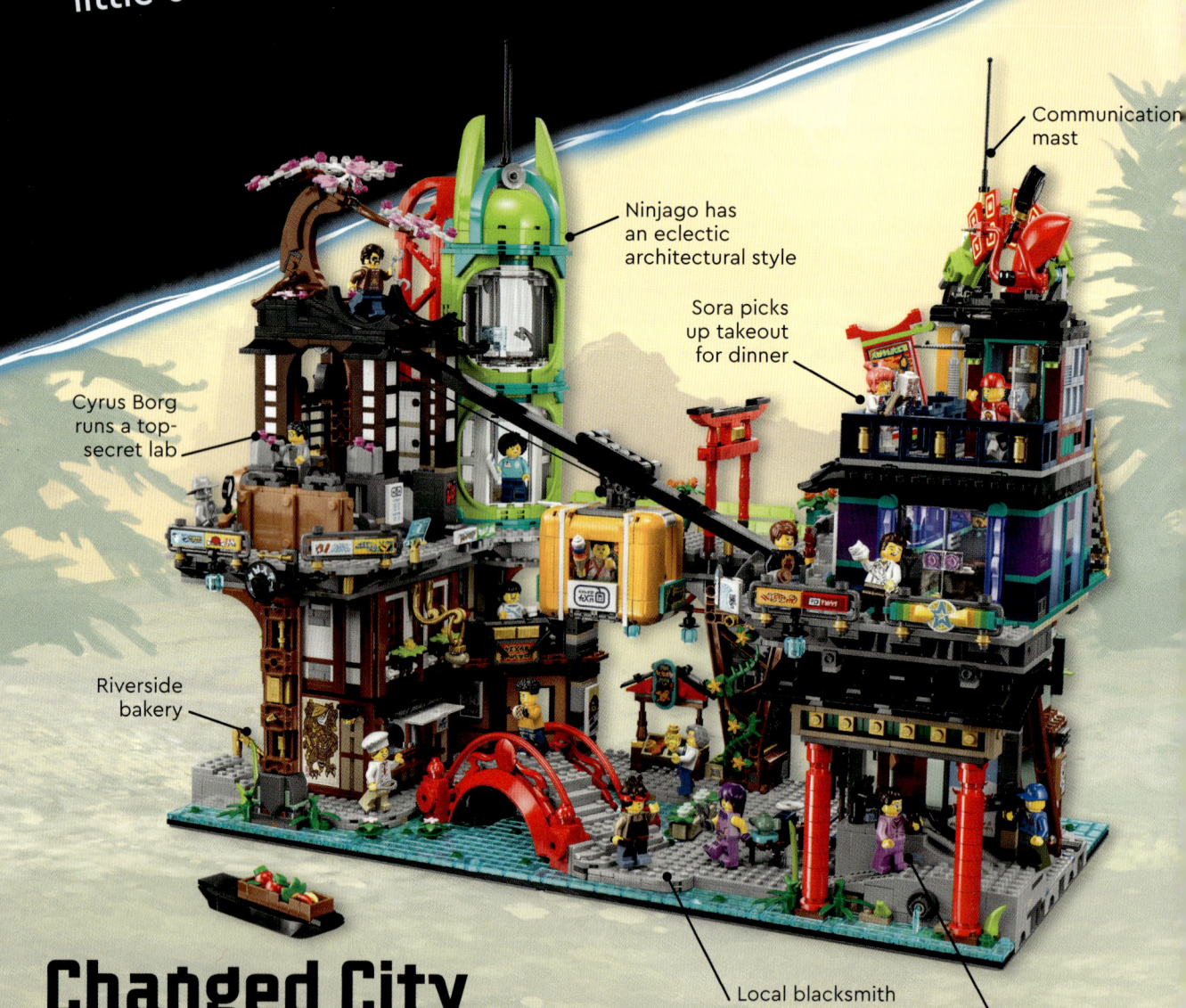

- Communication mast
- Ninjago has an eclectic architectural style
- Sora picks up takeout for dinner
- Cyrus Borg runs a top-secret lab
- Riverside bakery
- Local blacksmith Kai greets another Ninjago resident, Chamille
- Gayle Gossip reports the Ninjago news

## Changed City

Ninjago has a cable car, street market, and sushi place. But not everything is as it seems. Strange things happen here—realities shift, memories change, and new people appear.

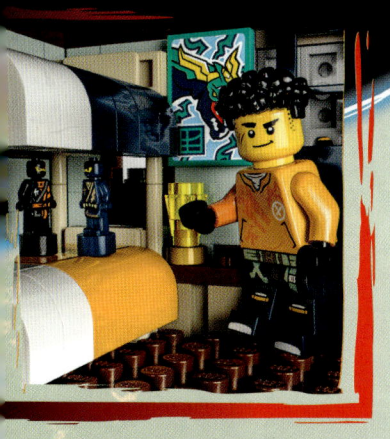

## Daydreaming

Arin dreams of being a Ninja. He loves telling stories of the old Ninja and their victories, and he plans to open his own Monastery of Spinjitzu one day!

*RIYU, HOW DID I DO THAT?*

## Surprise Powers

Sora is an inventor who lives with her friend Arin. One day, she meets a young dragon, Riyu, and a strange connection sparks between them. Sora is astonished to discover she may have powers!

## A New World

Everything is different from how it used to be. After the merge, which caused the realms to unite into one, the Ninja team split up. New heroes rise and the old heroes need to find their roles in this merged world.

*I'M GONNA NEED A BIGGER TEAM.*

## New Threats

A high-tech kingdom called the Imperium sends its army across the realms. Led by the imposing Empress Beatrix, the Imperium is powerful and dangerous.

## Lone Master

Lloyd has been traveling alone, searching for other Ninja, but he hasn't found any so far. When he meets Arin and Sora, he wonders if things are about to change.

# Can New Heroes Join Team Ninja?

**TIMES ARE CHANGING** and the Ninja pride themselves on how well they adapt to face new challenges. Team Ninja has evolved over the years, and when they meet two special new heroes, the Ninja are happy to welcome them to the team!

**TEAM NINJA WELCOMES 2 NEW MEMBERS!**

> RULE NUMBER ONE: NO PRANKS!

Nya has been learning all about dragons

## Water Welcome

Nya was the last Ninja to join the team, and she's happy to share her training tips.

Vehicle can perform special tricks

> WELCOME TO TEAM NINJA!

## Ice to Meet You

Zane is pleased to welcome his new teammates. He bets they've never met a Nindroid before.

Icicles leave no doubt about whose car this is!

# Arin

Arin has no elemental powers, but he has developed his own version of Spinjitzu—much to the Ninja's amazement. Imagine what he'll be like after some Ninja training!

Training mech has captured the villain Rapton

# Green Master

Lloyd is ready to train his new students, though he's still not used to being called Master Lloyd!

# Sora

Sora's strange powers seem to be connected to technology. She is grateful to Lloyd, who guides her on her journey to unlock her True Potential.

NINJA TRAINING? SOUNDS FUN!

Sora rearranged her mech into a jet, and then into this bike!

# Glossary

**AIRJITZU**
A variant of Spinjitzu that allows the user to hover or fly.

**AVATAR**
The version of a player inside a video game.

**CENTAUR**
A creature that is part human, part horse.

**CYBORG**
A living being that has robotic parts.

**DOJO**
A training ground specifically designed for martial arts practice.

**DRAGON FORM**
A higher form of elemental mastery that can only be unlocked with specific actions.

**DRAGON MASTER**
The title given to the person worthy of wearing the ancient Dragon Armor.

**DRAGONIDE**
Stone dragons with Vengestone wings that serve the Crystal King.

**ELEMENTAL MASTER**
An individual who wields a specific elemental power, which has been passed down through multiple generations.

**ENDLESS SEA**
The waters that surround Ninjago.

**FIRST SPINJITZU MASTER**
A fearless warrior who created Ninjago, and banished the Overlord to the Dark Island.

**FIRSTBOURNE**
The first dragon that ever existed and the source of all other dragons.

**GOLDEN MASTER**
A figure from Serpentine legend believed to bring darkness to Ninjago.

**GOLDEN NINJA**
The form Lloyd takes when he unlocks his True Potential as the Ultimate Spinjitzu Master.

**GREAT DEVOURER**
The snake that bit Lord Garmadon as a child. The Serpentine later unleashed the snake on Ninjago City.

**GREEN NINJA**
The strongest of all Ninja, destined to defeat the Dark Lord.

**HACK**
To access data in a computer or computer program in order to make changes.

**ISLAND OF DARKNESS**
An island that used to be part of Ninjago. It was split to keep the Overlord away from Ninjago, and it later sank beneath the sea.

**MECH**
A large, mobile suit of mechanized armor.

**NEVER-REALM**
The most remote realm in existence, rumored to be trapped in an endless winter.

**ONI**
Ancient demons with the power of Destruction, the Oni are the oldest evil in existence.

**REALM**
World or kingdom.

**SAMURAI**
A powerful warrior trained in martial arts and able to wield a variety of weapons.

**SERPENTINE**
An ancient race of snakes that has long inhabited Ninjago. They are divided into five distinct tribes.

**SPINJITZU**
A martial arts technique requiring the user to channel their elemental powers while spinning rapidly.

**SWORD OF SOULS**
A mystical blade that can trap souls within it.

**TORNADO OF CREATION**
A whirlwind of energy with the power to create something new from its surroundings or own elemental power.

**TRUE POTENTIAL**
The highest level of power a Ninja can unlock, achieved by overcoming personal challenges.

**UNDERWORLD**
A realm of Ninjago inhabited by the Skulkin.

**VENGESTONE**
A substance that cancels out elemental powers.

# Index

Main entries are in **bold**

## A
Anacondrai 24, **41**, 55
Arin 91, 93

## C
Cole 44, 45, 46, **48–49**, 50, 52, 63, 72, 78, 80
Crystal King 10, 49, 73, 84–85, 86, 88, 89

## D
Dark Island 33, 39
Dark Matter 15, 39
Destiny's Bounty 22
Dojo 11, 19, 23, 63, 66,
Dragons 27, **28–29**, 30–31, 35, 39, 47, 58, 59, 73, 78, 79, 81, 87, 89, 91, 92

## E
Elemental blade 46
Elemental Masters 11, 13, 24, **52–53**, 54
Elemental Power 9, 12, 17, 19, 27, 28, 31, 35, 41, 42, 46, 49, **52–53**, 55, 61, 73, 85, 93

## F
First Spinjitzu Master 7, **32**, 82
Firstbourne 6, 25

## G
Garmadon, Lloyd 13, 14, 17, 18, 22, 24, **26–27**, 28, 30, 32, 47, 52, 56, 57, 58, 62, 69, 79, 81, 86, 91, 93
Garmadon, Lord **14–15**, 16, 27, 30, 33, 41, 56, 69, 70, 76, 82
Garmadon, Sons of 25, 41, 69, **70**, 87
Garmatron 15
Golden Dragon 11, 21, 26, 31, 43
Golden Master 38
Golden Ninja 18, 26
  see also Garmadon, Lloyd
Golden Weapons 10, 24, 33, 38, **47**, 73, 76, 83, 85, 89
Great Devourer, the 15, 16, 32, 50, 68, 69, 76
Green Ninja 17, 18, 27, 57
  see also Garmadon, Lloyd

## H
Harumi, Princess **68–69**, 84

## J
Jay **20–21**, 35, 47, 50, 52, 63, 73, 81

## K
Kai **10–11**, 19, 29, 35, 37, 45, 46, 47, 52, 72, 75, 81, 83, 90

## M
Master Chen 24, 55, 60, 61
Master Chen's Island 23, 24
Misako 86
Morro 24, 57

## N
Nindroids 40
Nunchucks of Lightning 47, 88
Nya 11, 18, 20, 21, 23, 29, 32, **34–35**, 36, 47, 52, 80, 92

## O
Oni 10, 15, 25, 27, 33, 41, 71
Overlord, Digital 64
Overlord, the 15, 26, 33, **38–39**, 40

## P
P.I.X.A.L. 11, 42, 43, **64–65**, 73
Pythor 9, 84

## R
Riyu 91

## S
Samukai 24
Samurai X 18, 34, 65
  see also Nya
Scythe of Quakes 46, 85,
Serpentine 40, 41
Skulkin 24, 76
Skylor 9, 11, 53, **60–61**, 87
Sora 90, 91, 93
Spinjitzu 7, 14, 18, 19, 23, 26, 32, 33, 48, 66, 67, 82, 83, 91, 93
Stone Army 39
Sword of Fire 47, 83

## T
Titanium Ninja 42
  see also Zane
Tournament of Elements **54–55**, 61
True Potential 28, 34, 93

## U
Underworld 24, 76

## V
Vengestone 11, 43, 68, 84, 85, 88, 89
Vengestone Army 49, 84, 85, 86, 87

## W
Wu **6–7**, 8, 12, 13, 14, 19, 23, 24, 25, 27, 30, 32, 33, 56

## Z
Zane 19, 25, 33, 40, **42–43**, 47, 52, 57, 58, 65, 73, 81, 83, 92

A NINJA NEVER STOPS LEARNING!

**SENIOR ART EDITOR** Lauren Adams
**DESIGNER** Thelma-Jane Robb
**PROJECT EDITOR** Lisa Stock
**EDITOR** Nicole Reynolds
**SENIOR US EDITOR** Megan Douglass
**PRODUCTION EDITOR** Siu Yin Chan
**SENIOR PRODUCTION CONTROLLER** Lloyd Robertson
**MANAGING EDITOR** Paula Regan
**MANAGING ART EDITOR** Jo Connor
**PUBLISHING DIRECTOR** Mark Searle

Additional photography by Gary Ombler

DK would like to thank:
Randi K. Sørensen, Heidi K. Jensen, Martin Leighton Lindhardt, Morten Rygaard Johansen, and Tommy Kalmar at the LEGO Group; author Shari Last and proofreader Julia March.

First American Edition, 2023
Published in the United States by DK Publishing
1745 Broadway, 20th Floor, New York, NY 10019

Page design copyright © 2023 Dorling Kindersley Limited
A Penguin Random House Company

This book was made with Forest Stewardship Council™ certified paper—one small step in DK's commitment to a sustainable future. For more information go to www.dk.com/our-green-pledge

Contains content previously published in LEGO® NINJAGO® Secret World of the Ninja (2015)

www.LEGO.com/ninjago

24 25 26 27 10 9 8 7 6 5 4 3
007–336380–October/2023

LEGO, the LEGO logo, the Minifigure, the Brick and Knob configurations, and NINJAGO are trademarks of the LEGO Group. ©2023 The LEGO Group. Manufactured by Dorling Kindersley, One Embassy Gardens, 8 Viaduct Gardens, London SW11 7BW under license from the LEGO Group.

Without limiting the rights under the copyright reserved above, no part of this publication may be reproduced, stored in or introduced into a retrieval system, or transmitted, in any form, or by any means (electronic, mechanical, photocopying, recording, or otherwise), without the prior written permission of the copyright owner.

A catalog record for this book is available from the Library of Congress.

ISBN: 978-0-7440-8463-4
ISBN: 978-0-7440-8464-1 (Library Edition)

Printed and bound in China

For the curious
www.dk.com
www.LEGO.com